THE MATH TUTOR'S FRIEND

probability

scatterplot

misleading statistics

percentile

pie chart

z-score

BOOK 5

PROBABILITY & STATISTICS

by
Joe Cyr

Cyr, Joseph

The Math Tutor's Friend / Book 5 – Probability & Statistics / by Joe Cyr

Third Edition ISBN: 978-1-967039-46-3 Joseph Cyr

Second Edition ISBN: 979-8-9879759-5-4 Joseph Cyr

First Edition ISBN: 9798489136747 Amazon Kindle, 2022

3 4 5 6 7 8 9

Subjects: Main category of book – *counting; probability; statistics; data display* – high-school math

2.Other category - math tutor reference

Cover illustrations by Ramon E. Owen. Cover Design by Joe Cyr

Book design by Joe Cyr

The Math Tutor's Friend Series

Books

1 - Pre-algebra
2 - Algebra I
3 - Algebra II
4 - Algebra III
5 - Probability & Statistics
6 - Geometry
7 - Trigonometry
8 - Pre-calculus
9 - Series index & "FLASHMATH" handouts/reference cards

THE MATH TUTOR'S FRIEND

TABLE OF CONTENTS

BOOK 5

PROBABILITY & STATISTICS

Sections

About the "Math Tutor's Friend" Series:

Teachers impart knowledge guided by lesson plans in a course syllabus. Math tutors, on the other hand, face varying topics at differing levels in successive sessions; for example, pre-algebra, advanced algebra, geometry, trigonometry, statistics, or even calculus! This series addresses the issue and is intended to provide the tutor with needed quick review of high-school math topics. The material is not presented rigorously nor does it abound with proofs and explanation that are available in textbooks - the tutor already knows this stuff - but a quick refresher is always a good move before diving into a tutoring session.

Too many in high-school have reached the 11th and 12th grades lacking basic math skills such as knowing the multiplication table (relying on a calculator to multiply 6 by 7, for example), handling signed numbers, and performing operations involving fractions - knowledge that should have been acquired by the fifth grade! I try to address this deficiency by devoting the first few tutoring sessions to these fundamentals and providing handouts of multiplication tables,etc.

Books 1 through 8 of this series include a number of quiz questions (answers explained) that the math tutor may find useful.

Book 9 is an overall index of topics. It contains a set of summary "FLASHMATH" pages for quick reference or for use as handouts.

This third edition corrects typos found in previous editions and expands discussion of some of the topics.

Happy Tutoring!

Joe Cyr

Sarasota, FL

THE MATH TUTOR'S FRIEND
odds
probability tree
permutations
It's easy !!
combinations
probability
outcome
Mr. Math
COUNTING
&
PROBABILITY

About Book 5: "Probability & Statistics"

Section 1, *Counting & Probability*.

Emphasis -:

- Factorials
- Combinations & permutations
- Outcomes vs events
- Probability and odds
- Probability with and without replacement
- Conditional probability

Section 2, *Basic Statistics*.

Emphasis -:

- Types of studies
- Central tendency and outliers
- Measures of spread
- Properties of a probability distribution
- Z-score and percentiles
- Box-and-whisker plot

Section 3, *Organizing & Presenting Data*.

Emphasis -:

- Two-way frequency tables
- Plotting data
- Types of plots and graphs

• •

1. COUNTING & PROBABILITY

TABLE OF CONTENTS

•

1. COUNTING & PROBABILITY

Factorial

The *factorial* of a positive integer "*n*" is denoted by " *n*! " which is the product of all positive integers less than or equal to *n*:

$$n! = n\times(n-1)\times(n-2)\times(n-3)\times\ldots\times(n-(n-1)).$$

Example: $6! = 6\times5\times4\times3\times2\times1 = 720$

Zero factorial is defined to be one; that is, $0! = 1$ (Note: $1! = 1$ also).

One use of the factorial is to calculate permutations and combinations.

Counting – the simple case

If there are *m* ways to do one thing, and for each of these ways you can do a second thing in *n* ways, then you can do both in $m\times n$ ways.

Example: A four-course menu offers 2 kinds of soup, 3 kinds of salads, 4 entrees, and 5 desserts. How many meals can you order without having the same four-course meal twice?

• •: 2 soups×3 salads×4 entrees×5 desserts = $2\cdot3\cdot4\cdot5 = 120$ unique meals.

Consider the following: flip a coin and roll a pair of dice. The possible outcomes are H,T (for the coin), 1, 2, 3, 4, 5, 6 (for die #1), and 1, 2, 3, 4, 5, 6 (for die #2). There are two possible outcomes for the coin and six possible outcomes for each of the dice, so there are $2\times6\times6 = 72$ possible outcomes (H,1,1; H,1,2; etc.)

Example: A quiz has one True-False question and five multiple-choice questions, each having five selections. If you answer each question and leave none of them blank, in how many ways can you answer the whole test?

• • There are $2\times5\times5\times5\times5\times5 = 2\times5^5$ or 6,250 possible answers.

Combinations

The notations ${}_nC_r$, *C(n, r)*, ${}^nC_r$ and $\binom{n}{r}$ indicate the "*combination* of *n* things *r* at a time."

Example: Consider a set of eight baseball caps of different colors: black; yellow; green; purple; orange, white; turquoise; and red. Identify them as B, Y, G, P, O, W, T, and R, respectively. How many combinations of four of these are possible? • • For example, YGBR and GTRO are two combinations. Order is not important, so GYRB is the same as GYBR. Let *n* be the number of caps ($n = 8$ in this example) and *r* be the number that we choose ($r = 4$ in this example). The number of combinations of *n* things taken *r* at a time, denoted by ${}_nC_r$, is given by the formula $n!/[r!(n-r)!]$.

For the example we want to compute ${}_8C_4 = 8!/4!(8-4)! = 70$, so there are 70 combinations of four hats from a group of eight hats of different colors.

Counting subsets.

How many subsets are there in a set of *n* objects?

Example: A high school offer four electives. A student may take any number, from none to all of 4 them. How many different elective programs are possible?

• • The possible programs are $_4C_0$ (= 1), $_4C_1$(= 4), $_4C_2$(= 6), $_4C_3$ (= 4), $_4C_4$ (= 1). The total is 16, which is 2^4. A set of *n* objects has exactly 2^n subsets. This number includes the *empty set*, ø, and the set itself. The number of *proper subsets* (which does not include the empty set or the set itself) is given by $2^n - 2$.

Permutations (arrangements)

A *permutation* is an ordered arrangement of the elements of a set. For example, 321 and 213 are two permutations of numbers in the set{1,2,3}. The number of permutations (arrangements) of *n* things *m* at a time, denoted by $_nP_m$, is given by the formula

$$_nP_m = n!/(n - m)!$$

Example: There are six permutations of three blocks labeled A, B, C, respectively: ABC, ACB, BAC, BCA, CAB, CBA. Using the formula:

$$_3P_3 = 3!/(3 - 3)! = 3!/0! = (3\cdot 2\cdot 1)/1 = 6.$$

Note that if $n = m$, ("*n* things *n* at a time") the number of permutations is simply $n!$ (or $m!$).

Distinguishable Permutations

For certain sets we need to be able to calculate *distinguishable permutations*. Consider the five-letter word "TREES." Permutations calculated in the usual way will produce $_5P_5 = 5! = 120$. However, some of the permutations would be identical, for example "EESRT" and "EESRT," because there are two indistinguishable letters (the "E's").

Example: "TALLAHASSEE," which contains three pairs of letters (LL, SS, EE) as well as three "A"s. Some of the permutations are indistinguishable from one or more of the others.

For a set of *n* objects of which n_1 are alike and one of a kind, n_2 are alike and one of a kind, $n_i!$ are alike and one of a kind, the number of distinguishable permutations is $n!/(n_1!\times n_2!\times \ldots \times n_i!)$

TALLAHASSEE, contains 11 letters; thus $n = 11$. Let $n_1 = 3$ be the number of "A"s, $n_2 = 2$ the number of "L"s, $n_3 = 2$ the number of "S"s, and $n_4 = 2$ the number of "E"s. The number of distinguishable permutations, then, is

$$11!/(3!\times 2!\times 2!\times 2!) = 11!/36 = 831{,}600.$$

In contrast, an 11-letter word in which all the letters are different has

$$11! = 39{,}916{,}800 \text{ permutations.}$$

Outcome and event

Outcome. An *outcome* is a possible result of an experiment or trial. Example: if you roll a die (which has six faces), the outcome will be 1, 2, 3, 4, 5, or 6. If you toss a coin, the outcome will be "Heads" or "Tails." The set of all possible outcomes is the *sample space*.

Event. An *event* may be denoted by a capital letter, for example, "B" is the event of selecting a black marble.

Difference between Outcome and Event

Outcome and *event* are not synonymous. An outcome is the result of a random experiment, like rolling a die has six possible outcomes. However, an "event" is a set of outcomes to which a probability is assigned. One possible event is "rolling a number greater than 7."

Probability and odds

Probability

Probability is a number between 0 and 1 inclusive (i.e., 0 and 1 are included) expressed as a decimal, a fraction, or a percentage. Denote the probability of event A occurring by "P(A)," and so $0 \le P(A) \le 1$. If $P(A) = 0$, then A won't happen; if $P(A) = 1$, then A will happen with certainty. So, the higher the probability of an event, the more likely it is that the event will occur.

> Example: Tossing a fair coin. Since the coin is fair, the two outcomes ("heads" and "tails") are both equally probable; the probability of "heads" equals the probability of "tails"; and since no other outcomes are possible, that probability is 1/2 (which could also be written as 0.5 or 50%).

Intuitive comments regarding probabilities. One frequently hears words such as "probable" and "likely," in the media. The following interesting citations gleaned from the internet may provide some insight (these are not definitions):

Highly probable – roughly 65 to 80%
Probable (event) – possibly true; roughly 50 to 65% [synonyms: credible; believable; plausible]
Improbable – roughly 20% [synonyms: far-fetched; incredible; implausible]
Likely – a greater-than-even chance of occurring; > 50%
Highly unlikely – roughly 10 – 20%

Note the "gaps:" 80 to 100% and 0 to 10%; shall we call them "Virtually certain" and "nil?"

From the above (very rough) estimates, we might order events (e.g., event A) as *certain* (that is, probability 1, or $P(A) = 1$), *highly probable*, *probable* (or *likely*), *improbable*, *highly unlikely*, and *impossible* (e.g., probability 0, or $P(A) = 0$).

Sets

A *set* is a collection of objects and can be expressed as a list such as "A = {g, 4, w, 2}" or pictorially with a *Venn diagram* such as that shown to the right.Consider two sets A and B. The *intersection* of A and B is written "A∩B;" it is the set of objects that are in both A and B. For example if A = {1, 2, 3, 4, 5, 6} and B = {1, 4, 6, 7, 8, 9} then A∩B = {1, 4, 6} because 1, 4, and 6 are contained in both sets. If there are no elements common to sets A and B, then the intersection of those sets is the *empty set*, ø. The *union* of sets is denoted by the symbol "∪." For this example A∪B = {1, 2, 3, 4, 5, 6, 7, 8, 9}; that is, A∪B consists of all of the elements of both sets.

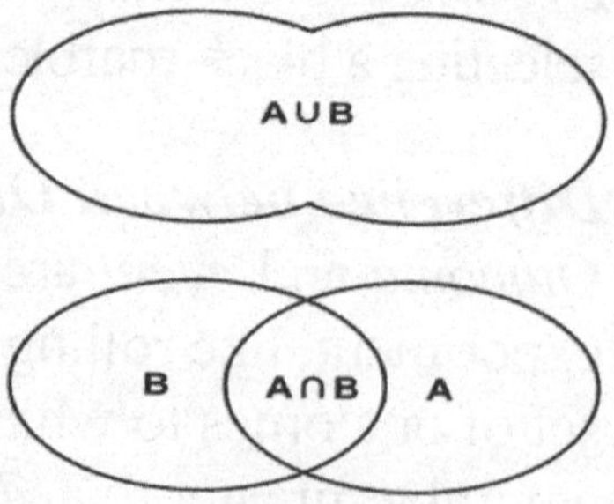

Now, let A be an event that occurs with probability P(A), and P(B) the probability of event B. The probabilities of intersection and union of those sets is denoted by

$$P(A \cap B) = P(A) \times P(B) \text{ and } P(A \cup B) = P(A) + P(B) \text{, respectively.}$$

Probability with replacement

Example: A sack contains 4 black marbles, 5 red marbles and 1 white marble. What is the probability of picking a red marble on the first try?

• • Let R be the event "select a red marble," B "select a black marble," and W "select a white marble." There are 10 marbles in the sack, 5 of which are red, so P(R), the probability of choosing a red marble on the first try, is (# of red marbles)/(Total # of marbles) = $5/10 = 1/2$ or 50%.

Now what is the probability of getting a black marble and two red marbles if we replace whatever we pick before making the next try? The fact that we replace a selected marble means that each try will be an independent event (not dependent on previous selections). P(B), the probability of picking a black marble is $2/5$ ($4/10$) and P(R) is $1/2$ ($5/10$). The probability of picking one black and two red marbles is the product of each independent event:

$$P(B) \times P(R) \times P(R) = 2/5 \times 1/2 \times 1/2 = 2/20 = 1/10 \text{ or } 10\%.$$

Probability without replacement

Example: A sack of marbles contains 4 black marbles, 5 red marbles and 1 white marble. Three marbles are selected. What is the probability of picking three red marbles?

• • Let R_1, R_2, and R_3 be the events "select a red marble on each of the 1st, 2nd, and 3rd tries. There are 10 marbles in the sack, 5 of which are red, so $P(R_1)$, the probability of choosing a red marble on the first try, is $5/10 = 1/2$ (50%). Now, only nine marbles remain, so, given that R_1 was successful, $P(R_2) = 4/9$, and given a success for R_2, $P(R_3) = 3/8$ because there are only three red marbles remaining in a sack of eight marbles. The probability, then, of picking three red marbles in a row is

$$P(R_1) \times P(R_2) \times P(R_3) = 1/2 \times 4/9 \times 3/8 = 12/144 = 1/12 = 8.3\%\text{, approximately.}$$

Conditional probability

The *conditional probability* of an event with respect to a prior event is the probability of occurrence of the event after the first event has taken place already and is denoted by P(B|A), the probability of event B, given event A has occurred.

Odds

The term "*odds*" (also "*odds ratio*") is related but not equivalent to the term "probability." If you say "The odds of winning are 20 to 5," then P_{win}, or P(win), the probability of winning, is ⅘ = 0.8 or 80%. The probability of losing is just 1 minus the probability of winning. Odds (odds ratio) are sometimes expressed as a *ratio of probabilities*, that is, the ratio of the probability of winning to that of losing: $P_{win}/(1 - P_{win})$. For the example, 0.8/(1 – 0.8) = 4:1, which is the same as "20 to 5" and "20⁄5 = 4." To convert from odds to a probability, divide the odds by one plus the odds. Thus P_{win} = 4/(1+4) = ⅘ = 0.8.

Difference between probability and odds

Example: A sack of colored marble contains ten balls: three white, five black, and two red. The probability, P(red), of selecting a red ball from this sack is 2⁄10 or 0.2 (20%) and the probability of NOT selecting a red ball is 1 – P(red).

Odds for is the ratio of occurrences of an event expressed to non-occurrences (*odds against*) of the same event. [$\boldsymbol{o}_f$(A) means "odds for A" and $\boldsymbol{o}_a$(A) means "odds against A" and $\boldsymbol{o}_f(A) = 1/\boldsymbol{o}_a(A)$; that is, *odds for* is the reciprocal of *odds against* : $\boldsymbol{o}_f = 1/\boldsymbol{o}_a$]. One notation is to use $\boldsymbol{o}_f$ ("odds for") and $\boldsymbol{o}_a$ ("odds against"). For example, the odds of picking a red ball is 2:8 because there are two red balls and there are eight balls in the sack which are not red. Because odds is a ratio, it is a fraction (as we learned in ratios and proportions), so the odds, $\boldsymbol{o}_f$ for selecting a red ball are 2⁄8 or 25%. Note that the odds against selecting a red ball , $\boldsymbol{o}_a$, are 8:2 or 400%, that is $1/\boldsymbol{o}_f$.

Convert from odds to probability and vice-versa

Let P(A) be the probability of event A occurring, and $\boldsymbol{o}_f$(A) be the odds that event A occurs. Then 1 – P(A) is the probability of event A *not* occurring and the odds against A occurring is the reciprocal of $\boldsymbol{o}_f(A) : \boldsymbol{o}_a(A) = 1/\boldsymbol{o}_f(A)$. If the odds of a horse winning are 3:12, the the odds of the horse losing (i.e., odds against the horse winning) are 12:3.

Using the sack of balls example, let r be the number of red balls in a sack of T (for total) balls. Then P = P(red), the probability of selecting a red ball is r/T. Now let S = $\boldsymbol{o}_f$(red) be the odds for the selecting of a red ball, so S is the ratio $r/(T - r)$. Solve the probability equation for r to get r = T×P(red) and substitute this in the equation for odds to get S = T×P(red)/(T – r) which simplifies to S = P(red)/(1 – r/T) = P(red)/(1 – P(red)). That is, the odds of an event occurring is the ratio of the probability of the event occurring to that of the event NOT occurring. So, we have expressed odds as a function of probability. Clearly, solving the equation for P(A) will express a probability of an event in terms of its odds. That is,

$$P(A) = \boldsymbol{o}_f(A)/(1 + \boldsymbol{o}_f(A))$$

and we have

$$\boldsymbol{o}_f(A) = P(A)/(1 - P(A)).$$

***Probability* (P) *vs Odds* ($\boldsymbol{o}_f$)**

(Impossible = $0 \leq P \leq 1$ = Certainty and $0 \leq \boldsymbol{o}_f < \infty$

$P < \boldsymbol{o}_f$ unless P= 0 (then $\boldsymbol{o}_f = 0$)

At values very near zero, $P \approx \boldsymbol{o}_f$; as P nears 1, $\boldsymbol{o}_f$ approaches ∞ .

Example: If the probability of winning is 40%, then the probability of losing is 60% (100% – 40%). The odds for winning are 40%/60% = ⅔ or 2 to 3, and the odds against winning are 3:2.

Probability Tree

The diagram above is a *probability tree* relating to a student completing a math sequence. At the start, 83% of the students will take an Algebra 1 class; given that Algebra 1 is completed, then 54% will opt for Algebra 2. Note that 38% of students complete only Algebra 1. Moving on through the tree we see that 64% of students completing pre-calc will take calculus, and of those, 90% will pass the course. Of interest are the "end points" of the tree. Note that the totals of the percentages sum* to 100%. In fact, the total of the branches from any point in the tree also sum to 100%. The numbers as discussed here are (hypothetically) historical, but the numbers could be viewed as probabilities. All of this stuff is called *combinatorics*.

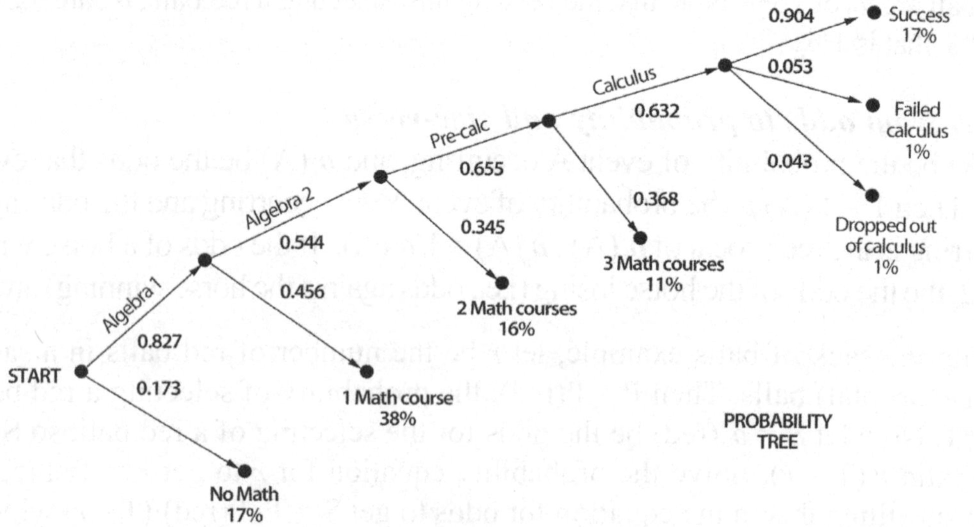

* Not quite, because of rounding errors for the decimals.

Tip regarding calculators

When using a calculator for combinations, factorials and permutations use $_nP_m$ instead of n!/(n – m)! because the numbers may become too large for the calculator.

Florida license plate example

The format for Florida Passenger plates is AAA A00 through ZZZ Z99.

Example: How many different six-character license plates are possible if the following restrictions apply: No "I," "O," or "Q" as first letter; no O, Q, or K as second letter; no I, O, K or Q as third letter. The fourth may be any letter, and the fifth and sixth characters may be any single digit 1 through 0.

$$23 \times 23 \times 22 \times 26 \times 10 \times 10 = 30{,}258{,}800 \ .$$

Types of Probability

Probability can be grouped as follows:

Theoretical probability – based on reasoning such as "If I roll a pair of dice then the probability of getting a three is $1/18$ because there are only two possible outcomes: "1" on the first die and "2" on the other (with probability $1/6 \times 1/6 = 1/36$); and "2" on the first die and "1" on the other (with probability $1/6 \times 1/6 = 1/36$). The probability of one or the other outcome is $1/36 + 1/36 = 1/18$.

Experimental probability – based on observations. If you roll a pair of dice ten times and get a seven four times, the experimental probability is $4/10 = 2/5$, even though the theoretical probability is $1/6$.

Axiomatic probability – involves formal statements and set theory and is beyond the scope of this book.

Handshaking Lemma

Imagine a political convention where many people shake hands. The *handshaking lemma** states that an even number of people must have shaken an odd number of other people's hands Take the simplified situation where four people shake hands. It takes two people to shake hands once. So person 1 greets person 2 and then person 3 and person four; that is, four people have shaken hands three times. So what? Well, that's another example of counting and it has many applications.

●

* A lemma is a "small" or minor theorem or proven proposition, often used in the proof of a major theorem.

SUMMARY OF PROBABILITIES

Source: Wikipedia

Summary of probabilities

Event	Probability
A	$P(A) \in [0,1]$
not A	$P(A^{\complement}) = 1 - P(A)$
A or B	$P(A \cup B) = P(A) + P(B) - P(A \cap B)$ $P(A \cup B) = P(A) + P(B)$ if A and B are mutually exclusive
A and B	$P(A \cap B) = P(A\|B)P(B) = P(B\|A)P(A)$ $P(A \cap B) = P(A)P(B)$ if A and B are independent
A given B	$P(A \mid B) = \frac{P(A \cap B)}{P(B)} = \frac{P(B\|A)P(A)}{P(B)}$

THE MATH TUTOR'S FRIEND
boxplot
percentile
median
It's easy !!
pictogaph
mean
mode
Mr. Math
BASIC STATISTICS

2. BASIC STATISTICS

TABLE OF CONTENTS

•

2. BASIC STATISTICS

Data set

Informally, a *data set* (or *dataset*) is a collection of data.

Example: A = {5, 13, 1, 2, 5, 8, 23, 2, 5, 6, 6, 10, 12, 11, 10} is a dataset containing 15 elements ($n = 15$).

Quantitative dataset -The set A above is a *quantitative dataset*; it consists of numerical values. A *qualitative dataset* consists of qualitative values or personal attributes ("tall," "short," etc.):

Example: B = {red, green, blue, purple, black, gray, olive}

Statistics

Statistics is the science of data. It involves collecting, classifying, summarizing, organizing, analyzing, and interpreting numerical information. A *statistic* is a calculated value (e.g., *mean* age of a group of individuals) relating to a *sample*; a *parameter* (or *data point* or *observation*) is a value relating to the *population* (e.g., age of an individual).

Types of statistics

Descriptive statistics utilize numerical and graphical methods to (1) look for patterns in data, (2) summarize the information revealed in a data set, and (3) present information in a convenient form

Inferential statistics utilize sample data to make estimates, decisions, predictions, or other generalizations about a larger set of data called the population.

Population. The entire group about which a characteristic is to be inferred.

Sample. A *sample group*) of a population is a *random sample,* i.e., a sample consisting of members of the group selected in a random manner. A *subgroup* is a smaller, distinct group within a sample, often defined by a specific characteristic. *Sampling* is employed when it is impossible to test every single individual in a population, and also to save time, money and effort relating to the research.

Census The ideal scenario is a census that tests every individual in a population to obtain reliable, valid and accurate results. While a census involves every member of a population, a *survey* (or *poll*) involves only the members of a sample group.

Surveys. A survey is a research method; people in a selected group are asked questions on different topics to acquire information. Surveys provide detailed information that is helpful in making predictions or decisions.

Polls. A poll, often used in politics, is a form of survey that usually involves a single question, perhaps multiple-choice or "yes/no," to a group of people concerning a person, subject, or some other issue. Results of polls can be announced immediately because there are usually few questions and little analysis performed on the data.

Types of studies

Observational studies and *experiments* are the two major types of study in research. The main difference between these is in the method of observation done.

Observational studies involves no human intervention. The researcher simply makes an observation and arrives at a conclusion. The researcher observes what has happened in the past and what is happening now and draws conclusions based on these data.

Experiments. An *experiment* entails human intervention. The researcher manipulates every aspect for deriving a conclusion. A *hypothesis* is tested by collecting and analyzing data to either accept or reject the hypothesis.

Analysis of data

Measures of central tendency

Consider the dataset A : A = {5, 13, 1, 2, 5, 8, 23, 2, 5, 6, 6, 10, 12, 11, 10}

A *central tendency* is a typical value near the middle of a (quantitative) dataset. The more common measures of central tendency are the *mean, median*, and *mode*:

Mean: The sum of all of the values in a dataset divided by the total number.

Example: The mean of dataset A is

(5 + 13 + 1 + 2 + 5 + 8 + 23 + 2 + 5 + 6 + 6 + 10 + 12 + 11 + 10)/15 = 7.9 (to one decimal point).

The mean is commonly called the "*average*."

Median: The middle value in a dataset arranged in ascending order:

Example: A = {1, 1, 3, 3, 5, 5, 5, 6, 6, 8, 10, 10, 11, 12, 23}

The middle value of 15 elements in the 8th number, 6, so the median of dataset A is 6.

NOTE: The median is the preferred measure of central tendency when the dataset contains very large and/or very small values. For example, in the case of salaries, where most of the amounts are from $8,000 to $130,000, say, but there are several cases of multi-million dollar salaries (baseball players?). The mean would pull the "average" to a much higher value that is not representative at all; therefore, the median would be the "average" of choice.

Example: suppose the dataset A (defined in the discussion of the mean) contained the additional value of 102:

A_1 = {5, 13, 1, 2, 5, 8, 23, 2, 5, 6, 6, 10, 12, 11, 10, 102}

Dataset A_1 has mean 13.8 (compare to 7.9 for A) and median 7 (compare to 6 for A).

When a dataset contains an even number of elements, the "middle" value is the average of the two numbers in the middle of the set.

Mode: The most frequently-occurring value in a dataset. The number "5" occurs most frequently in both example datasets, so the mode is 5.

Other measures of central tendency, such as the *geometric mean* and the *trimean,* exist. These are beyond the scope of this book.

Measures of spread

Range. The *range* of a set of data is the difference between the largest and the smallest values. The range of dataset A is $23 - 1 = 22$, of dataset A_1 is $102 - 1 = 101$. The range is sensitive to extreme scores since it is based on only two values.

$$A = \{5, 13, 1, 2, 5, 8, 23, 2, 5, 6, 6, 10, 12, 11, 10\}$$

$$A_1 = \{5, 13, 1, 2, 5, 8, 23, 2, 5, 6, 6, 10, 12, 11, 10, 102\}$$

Standard Deviation. *Standard deviation* is a measure of how spread-out the data in a sample are, and how close individual data points are to the mean value of the sample. Determining the standard deviation is a tedious process involving the calculation of the difference of each data element from the mean, squaring, summing (to find variance) and then taking the square root to obtain the standard deviation. Calculators for computing the standard deviation for a dataset are available on the Internet. The standard deviations for datasets A and A_1 are 5.6 and 24.1, respectively.

Weighted mean

The *weighted mean*, or *weighted arithmetic mean*, or *weighted average*, is used often by teachers. It is simply the sum of the individual weights multiplied by the individual scores, all of this divided by the sum of the weights. where $\bar{x}$ is the weighted mean.

$$\bar{x} = \frac{\sum_{i=1}^{n} w_i x_i}{\sum_{i=1}^{n} w_i}$$

<u>Example</u>: Assigning a large weight to a final exam may encourage students to work to improve their overall grade by doing well on the final. A teacher assigns weights to the academic efforts: homework, "pop quizzes," mid-term exam, and final exam (also lab work, if applicable). The assignments are 15% for homework assignments, 15% for pop quizzes, 20% for a midterm, and 50% for the final exam.

Notice that the percentages add to 100%, but it is not necessary for the weights to be percentages or even to add to 100.

Example: A teacher assigns the following weights for class work:

20 for homework assignments,
25 for pop quizzes,
40 for a midterm, and
80 for the final exam;

The weights add up to 165. In addition, the teacher is generous and rounds scores below 70 up to the next "ten;" thus, scores of 51, 57, 53 get rounded up to 60. With a highest possible score of 100 and a passing score of 70, a student gets the following scores during the semester:

Homework assignments: 40, 50, 60, 50, 40, 30.
Quizzes: 50, 60, 20, 50, 40.
Midterm: 60.
Final exam: 97.

The mean homework score is $(40 + 50 + 60 + 50 + 40 + 30)/6 = 270/6 = 45$

The mean pop-quiz score is $(50 + 60 +20 + 50 + 40)/5 = 220/5 = 44$

The *weighted mean* is the semester grade for the course:

[20×(average homework score) + 25×(average pop-quiz score) + 40×(mid-term score) + 80×(final exam score)]/(20 + 25 + 40 + 80)

The weighted mean (rounded to the nearest integer) is

$$(20\times45 + 25\times44 + 40\times60 + 80\times97)/(20 + 25 + 40 + 80) = 12{,}160/165$$
$$= 73.69697 \approx 74$$

Because of nearly "acing" the final, the student "squeaked by" to pass the course. This despite having poor grades on homework assignments and quizzes and failing the midterm.

Probability distribution

A *probability distribution* is a statistical function that describes all the possible values and likelihoods that a *random variable* can take within a given range.

Z-Score

A *z-score* indicates how many standard deviations an observation, x, is from the mean of the distribution. Calculate the z-score by subtracting the mean, x, from the observed value, $\overline{x}$, and dividing that difference by **s**, the standard deviation.

$$z = (x - \overline{x})/s$$

For a *normal distribution* (the "bell-shaped curve") approximately 68% of the data are within one standard deviation from the mean, about 95.5% between two standard deviations, and 99.7% of the data lie between three standard deviations.

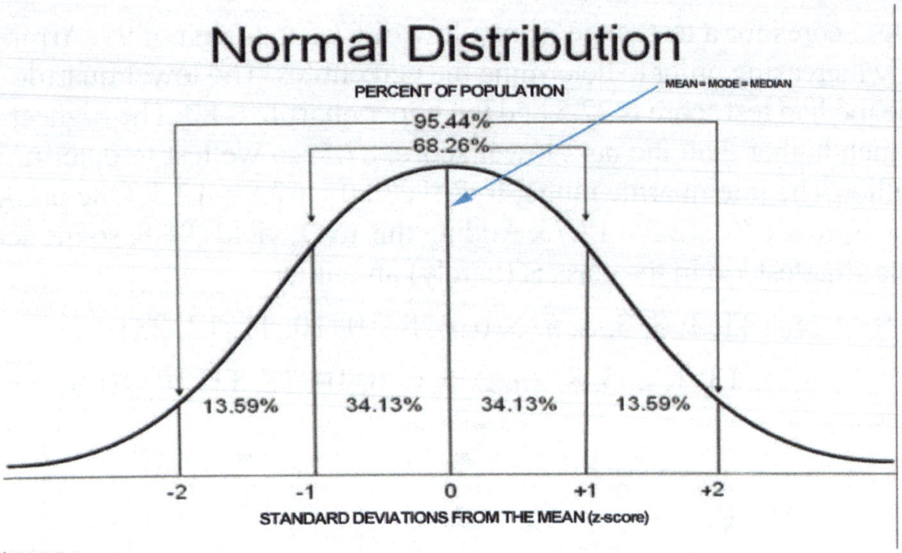

Percentiles

A *percentile* is a measure that indicates the value below which a given percentage of observations in a group of observations falls. For example, the 30th percentile is the value (or score) below which 30% of the observations may be found. We have already seen the 50th percentile: the median. The *lower quartile*, Q_1, is the "median of the median" and is the 25th percentile. The median, Q_2, is the 50th percentile, and Q_3, the 75th percentile is the *upper quartile.*

Interquartile range

Of special interest in the *interquartile range*, $IQR = Q_3 - Q_1$, a measure of how spread-out the values are. The IQR tells how spread out the "middle" values are; it can also be used to tell when some of the other values are "too far" from the central value. Interquartile range statistics assumes that your values cluster around some central value.

Outliers

Data values greater than $Q_3 + 1.5\times IQR$ and smaller than $Q_1 - 1.5\times IQR$ are *outliers.*

Box-and-whisker plot

A special graph that displays quartiles, IQR, the range and any outliers is the *box-and-whisker plot*, also called the *boxplot*, a concise graph summarizing the *five-point values*: lowest,; 1st quartile; median; 3rd quartile; highest. Multiple boxplots can be drawn side by side or "stacked" to compare more than one data set.

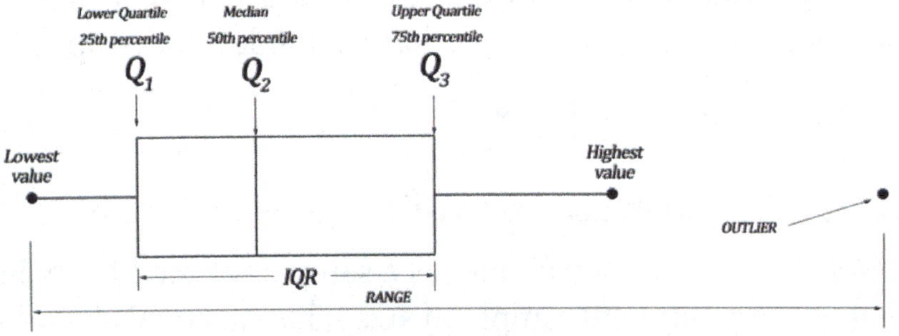

BOX & WHISKER PLOT NOMENCLATURE

Example: Scores for a test range from a low of 63.5 to a high of 99. Arrange the scores by increasing order to determine the percentiles. The lower quartile, Q_1, is 67.5; the median test score is 72.5 and the upper quartile is 80. The highest score, 99, is much higher than the next lower score, 87.5, so we test to determine if it is an outlier. The interquartile range, IQR, is 80.0 – 67.5 = 12.5. One-and-a-half times the IQR is 1.5 × 12.5 = 18.75. Adding this to Q_3 yields 98.8, so the score of 99 by the smartest kid in the class is (barely) an outlier

$$A = \{1, 1, 3, 3, 5, 5, 5, 6, 6, 8, 10, 10, 11, 12, 23\}$$

$$A_1 = \{5, 13, 1, 2, 5, 8, 23, 2, 5, 6, 6, 10, 12, 11, 10, 102\}$$

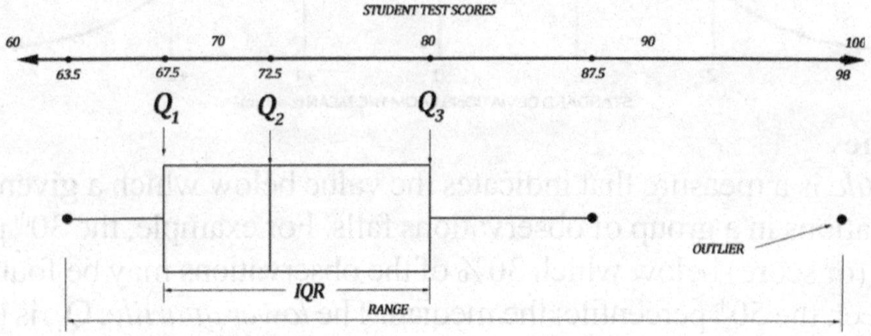

Skewness

Skewness is the measure of the asymmetry of a probability distribution. The skewness of dataset A is 1.42 and of dataset A_1, 3.22 (search the Internet for skewness calculators). A distribution is skewed if one of its tails is longer than the other. See the figure on the next page. Distributions with positive *skew* are more common than distributions with negative skews, especially when the smallest value of the domain is zero (i.e., no negative numbers) such as income.

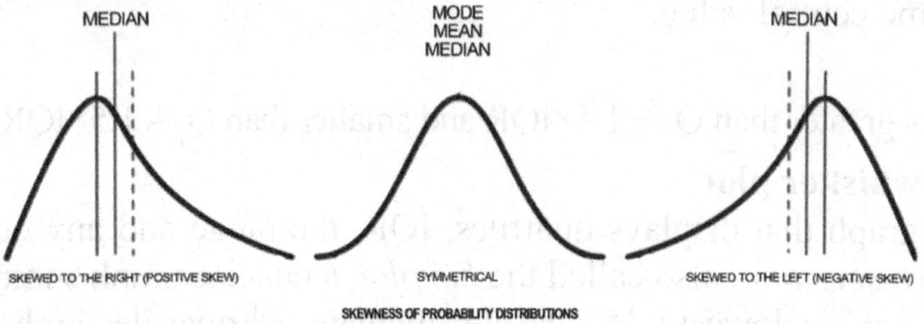

SKEWNESS OF PROBABILITY DISTRIBUTIONS

Most people earn less than $50,000 a year, but some make quite a bit more with a small number making many millions of dollars per year. The positive tail therefore extends out quite a long way whereas the left tail stops at zero.

Additional terms used in statistics

Samples and data

Representative samples exhibit characteristics typical to those of the population.

A *random sample* on n experimental units is a sample selected from the population in such a way that every different sample of size n has an equal chance of selection.

Per capita is a Latin phrase literally meaning "by heads" or "for each head", and idiomatically used to mean "per person". The term is used in a wide variety of social sciences and statistical research contexts, including government statistics, economic indicators, and environment studies. It is commonly and usually used in the field of statistics in place of saying "per person" [Source: *Wikipedia*]

Misleading statistics

Statistics can be misleading through ignorance of the user or malevolence of those producing the results, such as in advertising and politics. We hear phrases like "A recent study shows…," "The survey proves…," etc. One often sees questions on a survey phrased to influence the answer – you see much of this in political surveys and polls. So, statistics can mislead accidentally (poor techniques for gathering data) or deliberately, such as *data framing* (questions worded to influence respondents). Here are some examples of *misleading statistics*:

Faulty polling. Even if poll questions are fairly and neutrally stated, the respondents may not be a random sample. For example, polling on land-lines addresses only people having land-lines who answer the phone. Polling may take place only in a population with a majority of one political party, and so on.

Flawed correlation. It is possible that one could correlate the price of hamburgers in Alaska with the cost of college tuition in Florida. Correlation is NOT causation! One cynical report on the Web states that between 2000 and 2009, the per capita consumption of cheese in the U.S. correlates ($> 94\%$) with the number of people who died by becoming tangled in their bedsheets

Data fishing, also called *data dredging, data snooping, data butchery, p-hacking,* is the misuse of data analysis to find desired patterns and correlations Today's massive computer power and *artificial intelligence* (AI) allows data mining in all sorts of files to extract data. Enough *data fishing* could probably produce any desired result.

Selective bias (also *selection bias*) is the selection of individuals in a non-random way; the results could be biased to favor a particular point of view. In other words, the sample is not randomly selected. The timing of a poll could affect the results. Selection bias results when a subset of the experimental units in the population is excluded so that these units have no chance of being selected in the sample. There are many forms of selective bias.

Sampling bias prevents the sample from objectively representing the population under test. Findings from biased samples can only be generalized to populations that share characteristics with the sample. *Sampling bias* includes *self-selection, pre-screening* of trial participants, *discounting trial subjects/tests* that did not run to completion and *migration bias* by excluding subjects who have recently moved into or out of the study area.

Non-response bias results when the researchers conducting a survey or study are unable to obtain data on all experimental units selected for the sample.

Generating percentages from small sample sizes

Example: "50% of girls flunked math while only 25% of boys flunked the course." In actuality, more boys than girls flunked the course because one of the two girls in the course flunked while four boys (out of 16) flunked. One could also have said, accurately, that four times as many boys than girls flunked the course.

There are appropriate statistical methods and tests to deal with small sample sizes.

Errors in statistical analysis

False positive (also *false alarm, type I error*) is a result that indicates a given condition exists, when it does not.

Example: A woman is declared pregnant when she is not.

The false positive error is a type I error where the test of a single condition wrongly gives an affirmative (positive) decision - the error of rejecting a null hypothesis when it is actually true. In other words, this is the error of accepting an alternative hypothesis when the results can be attributed to chance. A Type I error occurs when we are observing a difference when in truth there is no statistically significant difference.*

False negative (also *type II error*) is a test result that indicates that a condition does not hold, while in fact it does. In other words, erroneously, no effect has been inferred.

Example: A guilty person is acquitted of a crime.

The *false negative* error occurs in a test where a single condition is checked for and the (erroneous) conclusion is that the condition is absent. It is the error of not rejecting a null hypothesis when the alternative hypothesis is true. That is the error of failing to accept a true alternative hypothesis, i.e., failing to observe a difference that exists.

Mnemonic: "FIP" (one "I") - False (Type I error) Positive.
"FINI" (two "Is") - False (Type II error) Negative

Measurement error in a survey, for example, refers to inaccuracies in the values of data recorded. In surveys, the error may be due to ambiguous or leading questions and the interviewer's influence on the respondent.

* The claim that a result from data generated by testing or experimentation is not likely to occur randomly or by chance but is instead likely to be attributable to a specific cause.

Random sampling in experiments

The method of choosing a sample is important when designing statistical experiments. Samples from populations must be truly random. Tests for randomness are available. It is not always easy to obtain a random sample from a given population as inadvertent biases and patterns tend to crop up. One way to choose a random sample is to use a table of random numbers. Sloppy techniques such as repeated use of the same part of such a table can violate randomness. Another technique is to use a random number generated by a computer* or calculator, say, by assigning each member of the population a unique identifying number.

> Example: Assume a population of 1,000; assign each individual a number from 000 to 999 then generate a 9-digit random number and compile from it a number by selecting, say, the 3rd, 5th, and 9th digit of that random number, each selection to be made after generating a new random number. An (ca 2025) Iphone displays a "Rand" button on its calculator. The first number is 0.92583776 and so I choose the individual with pre-assigned number 536. The next "Rand" is 0.104412615 so I select the person with pre-assigned number 415, and so on. I might choose to count the digits alternately from left then from the right-most digit of the displayed random number. It is important to decide in advance whatever rule to use, and be consistent throughout the process.

This technique is practical only for small populations. Other rules are used for large or infinite populations. Nowadays, of course, random number generation and sampling can be done by a computer app or through use of Artificial Intelligence (AI). There are many random sampling generators on the internet. All of this is termed statistical control: the use of statistical methods to monitor and control the quality of a process or experiment.

●

* Some computer apps produce "pseudo-random" numbers generated from known seeds so that experiments can be replicated using the same "random" numbers generated in previous experiments. This, of course, is a deterministic process but the generated numbers do pass the various tests for randomness.

THE MATH TUTOR'S FRIEND
2-way table
scatterplot
box-and-whisker plot
histogram
pie chart
pictograph
It's easy !!
Mr. Math
DISPLAYING & PRESENTING DATA

3. ORGANIZING & PRESENTING DATA

TABLE OF CONTENTS

•

3. DISPLAYING & PRESENTING DATA

Class and frequency

A ***class*** is a category into which *qualitative data* can be classified.

The ***class frequency*** is the number of observations in the data set of a particular class.

The ***class relative frequency*** is the class frequency divided by the total number of observations in the data set.

The ***class percentage*** is the class relative frequency multiplied by 100%.

Presenting statistical data – tables, graphs and plots

Tables

Two-way Frequency Tables Some displays - bar graphs, for example - usually display only one variable (one category). A way to organize and analyze statistical data involving two categories is through use of the *two-way frequency table.*

Example: A survey of 358 high-school seniors accepted for college asked if they were going to pursue science courses or liberal arts courses. The results are displayed below in a two-way frequency table.

WILL YOU PURSUE SCIENCE OR LIBERAL ARTS IN YOUR UPCOMING COLLEGE WORK?			
	SCIENCE	***ARTS***	***TOTAL***
GIRLS	55	133	188
BOYS	102	68	170
TOTALS	157	201	358

The information displayed in the two-way frequency table shows how many girls (188) and boys (170) responded to the survey and the number of each who selected science and arts. The data (55, 133, 102, and 68) are *joint frequencies*, and the sums of the rows and columns are *marginal frequencies*. The data shown can be further displayed in a two-way relative frequency table that allows us to determine the percentage of each "box" of the total:

The table below shows that 15% of those who chose science are girls compared to 29% of the total, who are boys, and in all, 44% of the students selected science.

Will you pursue science or liberal arts in your upcoming college courses?			
	SCIENCE	*ARTS*	*TOTAL*
GIRLS	55/358 = 15%	133/358 = 37%	188/358 = 53%
BOYS	102/358 = 29%	68/358 = 19%	170/358 = 47%
TOTALS	157/358 = 44%	201/358 = 56%	358/358 = 100%

Conditional-frequency-for-rows table

suppose that one needs to know the percentage of boys and girls who selected each of the course of study (science and arts). Construct a conditional relative frequency for each row as shown below.

Will you pursue science or liberal arts in your upcoming college courses?			
	SCIENCE	*ARTS*	*TOTAL*
GIRLS	55/188 = 29%	133/188 = 71%	188/188 = 100%
BOYS	102/170 = 60%	68/170 = 40%	170/170 = 100%
TOTALS	157/358 = 44%	201/358 = 56%	358/358 = 100%

From the construction of the conditional-frequency-for-rows table we see, for example, that 29% of the girls chose science compared to 60% of the boys.

Conditional-frequency-for-columns table

Finally, let's look at Science and Arts, i.e., the columns. Take each joint frequency (data value) as a percentage of its column total: The table shows that thirty-five percent of the science tract selection is by girls, and 34% of the arts tracts are boys.

Will you pursue science or liberal arts in your upcoming college courses?			
	SCIENCE	*ARTS*	*TOTAL*
GIRLS	55/157 = 35%	133/201 = 66%	188/358 = 53%
BOYS	102/157 = 65%	68/201 = 34%	170/358 = 47%
TOTALS	157/157 = 100%	201/201 = 100%	358/358 = 100%

Bottom Line. Instead of trying to remember all of the terms involved with these tables, make the various constructions (e.g., rows, columns, total) to answer the questions such as "How much of the arts tract was chosen by girls?" or "How many boys chose science?" or "What % of the total selections was science?" These questions are answered by looking at the responses as percentage of column totals, row totals, or grand totals, respectively.

Stem-and-leaf plot

Although not visually appealing, a *stem-and-leaf plot* (or *stemplot*) provides a concise representation of data. Let's present dataset A in a stem-and-leaf plot.

$$A = \{1, 1, 3, 3, 5, 5, 5, 6, 6, 8, 10, 10, 11, 12, 13, 23\}$$

The single-digit numbers have a "stem" of "0" (01, 03, 05, 06, 08). The right-most digit is the "leaf." Variations can be made to accommodate large numbers, negative numbers, decimals, etc.

STEM	LEAF
0	1 1 3 3 5 5 5 6 6 8
1	0 0 1 2 3
2	3

Back-to-back stem-and-leaf plot

As already seen, a stem-and-leaf plot displays data by splitting up each value in a dataset into a "stem" and a "leaf." Stemplots may be used to display values for two datasets in a single expanded *back-to-back stem-and-leaf plot.*

Example: Two datasets ***A*** and ***B*** contain the following data (ordered):

$$A = \{ 1, 2, 7, 10, 12, 13, 13, 13, 19, 23, 26, 32, 35, 39 \}$$

$$B = \{ 4, 6, 7, 7, 7, 12, 14, 15, 16, 21, 22, 25, 27, 27, 30, 32 \}$$

Construct the back-to-back stem-and-leaf plot for **A** and **B** by creating a single "stem" that represents the first digit in each value in both sets and "leaves" that on either side of the stem to hold the "tens" digit for each value. See the diagram below.

Insert the values of both sets ("tens" place) are on both sides of the "stem" in increasing order from the stem, as shown in the diagram.

The back-to-back stem plot display allows easy analysis and comparison of the values in both datasets.

A		B
7 2 1	0	4 6 7 7 7
9 3 3 3 2 0	1	2 4 5 6
6 3	2	1 2 5 7 7
9 5 2	3	0 2

The diagram shown above displays the *mode* (13 for set A, 7 for set B) and the *range* (39 – 1 = 38 for set A, 32 – 4 = 28 for set B) for the data In addition, the following are easily determined from the back-to-back stem-and-leaf plot:

Scores above (or below) a specified value.

- Values greater than 27 for Set A: 32, 35, 39
- Values greater than 27 for Set B: 30, 32

Instances of scores above (or below) a specified value.

- Number of values greater than 13 for Set A: six
- Number of values greater than 13 for Set B: ten

High and low score.

- Highest score and lowest score for Set A: 39 and 1, respectively
- Highest score and lowest score for Set B: 32 and 4, respectively

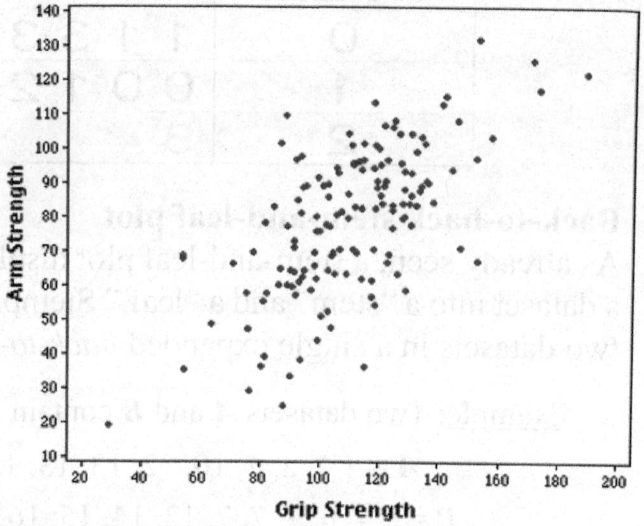

Scatterplots

A *scatterplot* displays the relationship between two variables of an experiment. A *trend line* is used to determine *correlation* between scores of one variable called the *dependent variable* plotted on the *y*-axis against scores on a second variable called the *independent variable* on the *x*-axis.

The scatterplot above shows the relationship between grip strength and arm strength for 147 people working at physically-demanding jobs. The data are from "a case study in the Rice Virtual Lab in Statistics." The plot shows that there is a linear relationship between these two variables.

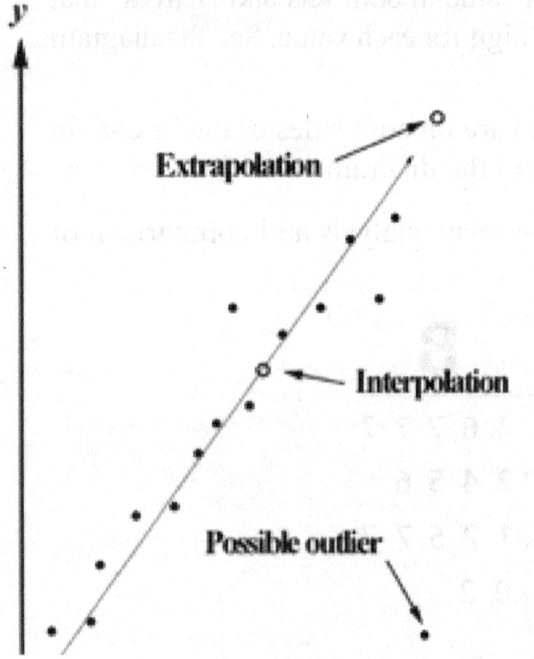

A sparser scatterplot (left) contains a superimposed trend line to allow *interpolation* for showing correlation of points not generated in the observations. It is possible to extrapolate using the trend line but this is not advisable; if the data curves, say, sharply downward, then the *extrapolation* (which is just a prediction) could be grossly incorrect.

Display of statistical data

Pictograph

A pictograph is a simplistic display that uses icons to represent. A key is used to explain the icon.

The pictograph on the left displays the (hypothetical) results of the status of students in various high-school courses of study. Each icon represents ten students; "smiley faces" are honor students, "neutral faces" are those who pass the course, "sad faces" those who failed the course, and "teary sad faces" dropped out of the course. So, for Civics/Current Events, for example, in a class size of 35 students, 10 were honors, 20 others passed, and 5 failed the course.

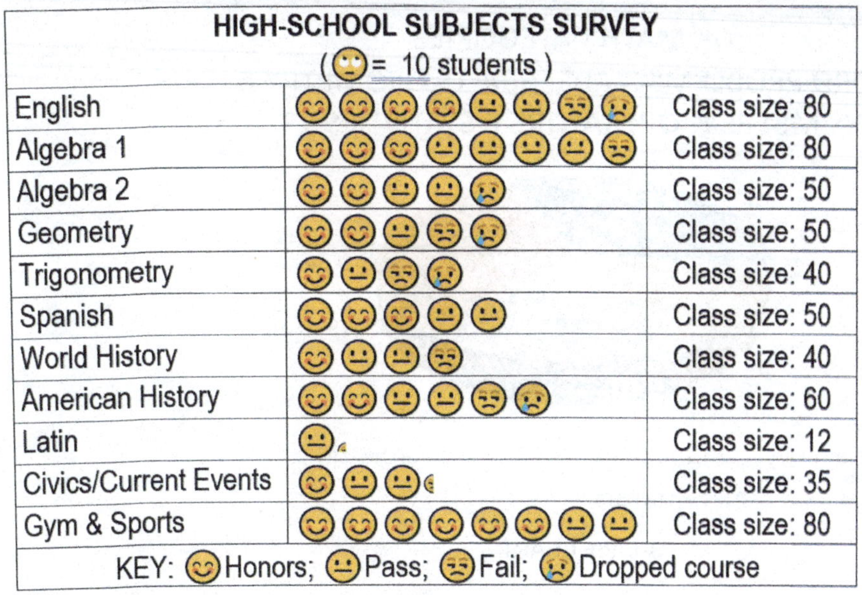

Line plot

A *line plot* is a way to display small (less than 50, say) amounts of data along a number line. It is easy to construct and requires no special tools or apps. Line plots are also called *dot plots*. A line plot displays range, minimum, maximum, modes and reveals gaps, if any.

Example: A hypothetical group of 38 high-school seniors were asked how much cash in $5, $10, or $20 bills they were carrying. The dot plot below shows the distribution of the 38 responses.

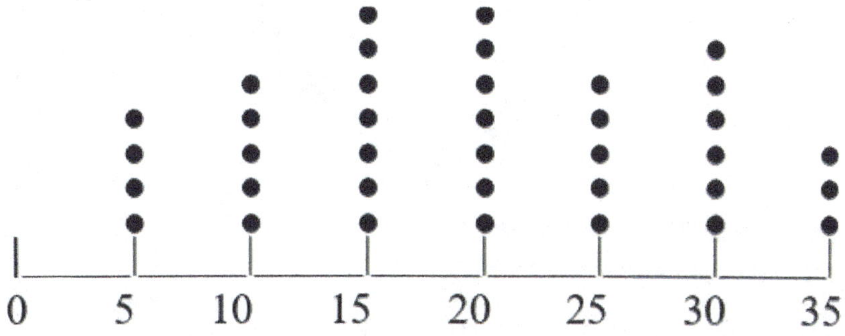

Pie Chart and Doughnut Chart

If there are not too many categories a visually appealing display is the Pie Chart and its cousin the Donut Chart. A *pie chart* displays data as a percentage of the whole, analogous to the area (or angle) of the pie slice. Each pie section should have a label and percentage. The pie chart and donut chart shown below display the relative emphasis of each GK math category.

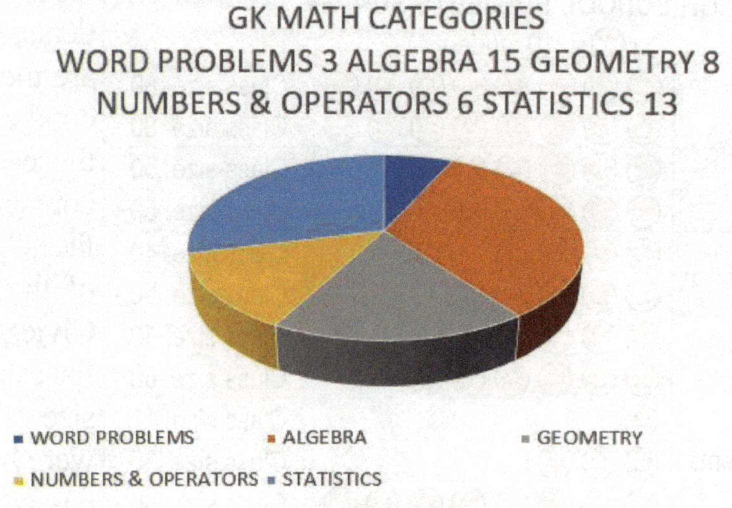

3-D PIE CHART

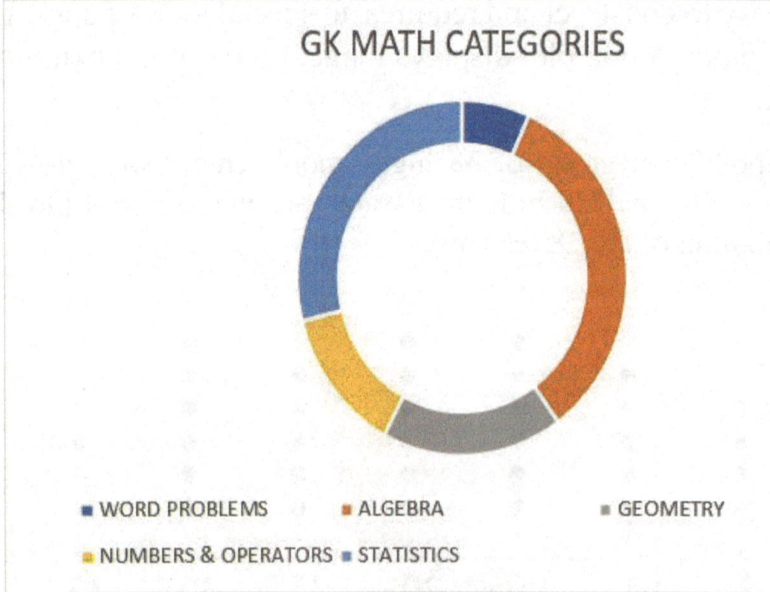

DONUT CHART

Line Graph

A *line graph* plots continuous data as points and then joins them with a line. Multiple data sets can be displayed together. The histogram chart shown below includes a line graph for showing cumulative data.

Histograms

A *histogram* displays continuous data in ordered columns. Categories are of continuous measure such as time, inches, temperature, etc. The histogram shown below displays the (hypothetical) number of questions in each General Knowledge (GK) Math category. The line graph showing cumulative numbers of questions is superimposed on the chart.

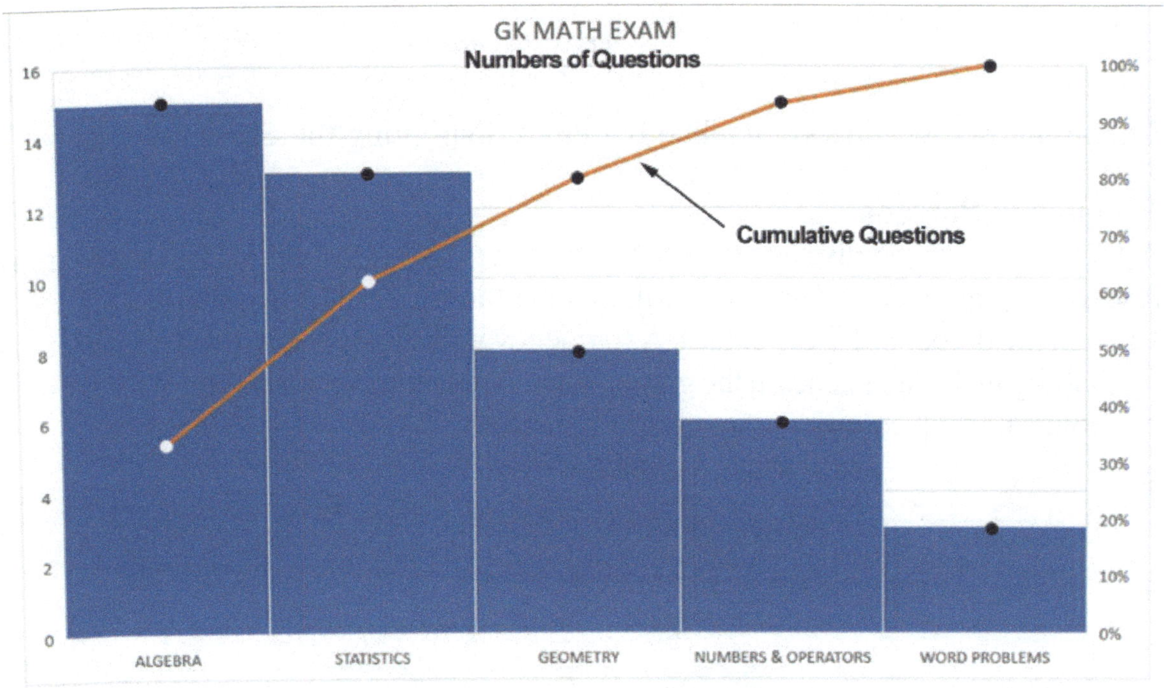

Bar Chart / Bar Graph

The *bar chart* is a popular method for displaying discrete data in separate columns. A double bar graph can be used to compare two data sets. The bar graph shown below is a (hypothetical) summary of the number of questions – easy and difficult – in each of five categories of GK Math.

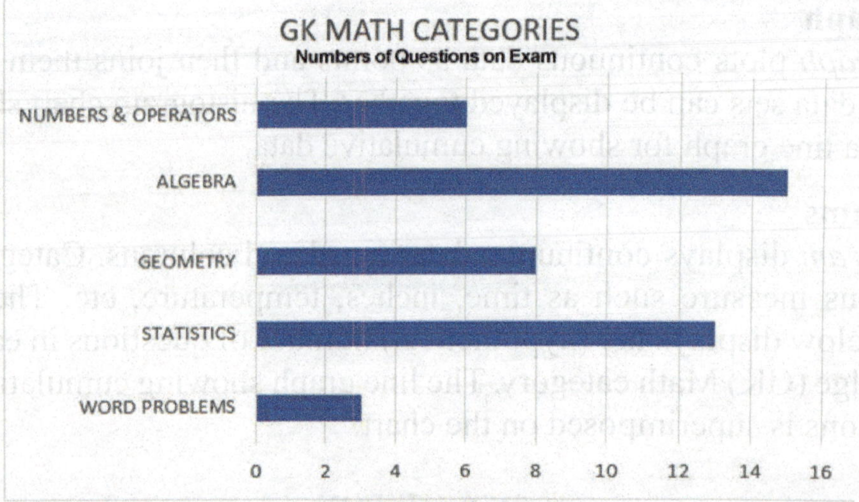

STACKED BAR CHART

The figure below shows a stacked bar chart displaying the exam questions.

Frequency Polygon

A *frequency polygon* displays the shape of a frequency distribution. The *x*-axis represents the values of a data set and the *y*-axis the number of occurrences of each value in the domain of the *x*-values. A frequency polygon is made from a line graph by shading in the area beneath the graph. It can be made from a histogram by joining midpoints of each column.

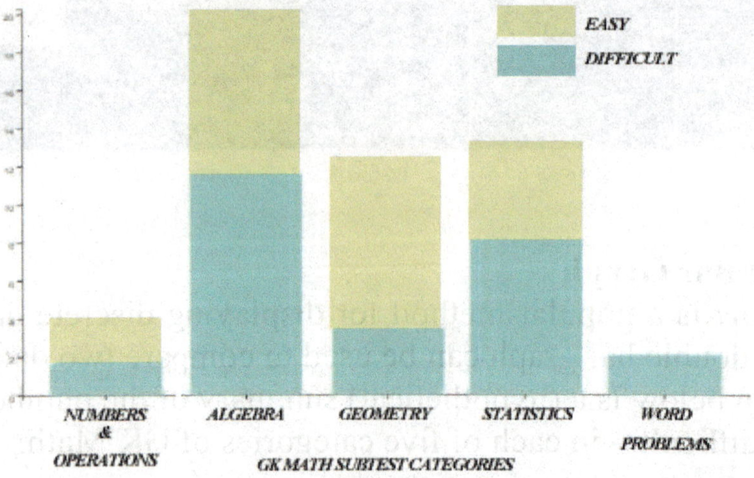

INDEX

Symbols

T

U

V

W

Z

• • •

TEST QUESTIONS

PROBABILITY & STATISTICS

QUIZZES

• • •

• The "five points" shown on the box of a boxplot are

A. Q_1, Q_2, Q_3, Q_4, Q_5

B. Low outlier, Q_1, Median, Q_2, high outlier

C. Median, Mode, Mean, Geometric Mean, Trimean

D. Smallest, Q_1, Q_2, Q_3, Highest

E. None of the above

• • Answer D.

The smallest value is the lower end of the left whisker;

Q_1 is the left edge of the box;

Q_2 is the Median and lies within the box (between Q_1 and Q_3)

Q_3 is the right edge of the box (opposite Q_1);

highest value is attached to the right whisker.

Outliers are not connected to the box.

• ♥ •

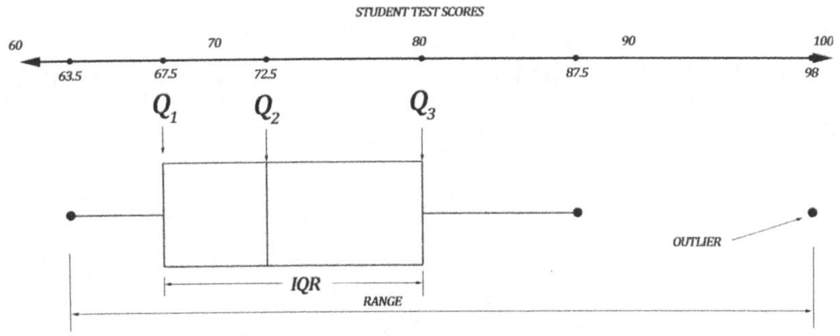

• The box-and-whisker plot shown above presents the distribution of student test scores. The median student test score is

A. 63.5

B. 67.5

C. 72.5

D. 80

E. 86.5

• • Answer C. The median is the 50th percentile, displayed as Q_2.

• A quiz has one True-False question and five multiple-choice questions, each having five selections. If you answer all questions and leave none of them blank, in how many ways can you answer the whole test?

A. 6

B. 25

C. 5,000

D. 6,250

E. None of the above

• • Answer D.

There are five events: questions 1 through 5, and there are five choices for each question.

There are $2 \times 5 \times 5 \times 5 \times 5 \times 5 = 2 \times 5^5 = 6{,}250$ possible answers.

• ♥ •

• How many outcomes are possible for a game involving flipping two quarters and rolling two dice?

A. 4

B. 144

C. 36

D. 72

E. 216

• • Answer B.

There are two outcomes for each coin and six outcomes for each die:

$$2 \times 2 \times 6 \times 6 = 144$$

• How many combinations of three bills can we select from a cash till containing \$1, \$2, \$5, \$10, and \$20 bills in separate compartments?

A. 5
B. 10
C. 15
D. 20
E. None of the above

• • Answer B.

$${}_5C_3 = 5!/[3!(5-3)!] = 10$$

• ♥ •

• There are _____ distinguishable permutations for the word "GEESE."

A. 120
B. 5
C. 20
D. 30
E. None of the above

• • Answer C.

The number of distinguishable permutations is $n!/(n_1!\times n_2!\times \ldots \times n_i!)$, here,

$$5!/3! = 5\cdot 4\cdot 3!/3! = 20 .$$

• How many possible arrangements (permutations) are there for a trifecta (1st place, 2nd place, 3rd place) in a horse race with ten entries?

A. 260

B. 120

C. 130

D. 720

E. None of the above

• • Answer D.

${}_{10}P_3 = 10!/(10 - 3)! = 10!/7! = 10{\cdot}9{\cdot}8{\cdot}7!/7! = 10{\cdot}9{\cdot}8 = 720$. Note that if order does not matter (e.g., 2, 5, 10 is the same as 5, 2, 10), then we look for the combination of 10 things 3 at a time:

$${}_{10}C_3 = 10!/[3!(10 - 3)!] = 10!/(3!{\cdot}7!) = 720/3! = 720/6 = 120$$

• ♥ •

• Roll two fair dice and toss two fair coins (H or T). What is the probability of

(1) Rolling two sixes ("midnight") and getting two Heads?

(2) Rolling two ones ("snake eyes") or getting two Tails?

A. (1) $\frac{1}{36}$; (2) $\frac{1}{2}$

B. (1) $\frac{1}{6}$; (2) $\frac{1}{4}$

C. (1) $\frac{1}{36}$; (2) $\frac{1}{4}$

D. (1) $\frac{1}{104}$; (2) $\frac{5}{13}$

E. None of the above

• • Answer D.

First note that for the dice, the probability of two sixes is the same as the probability of two ones, namely, $\frac{1}{6} \times \frac{1}{6} = \frac{1}{36}$. For the coins, The probability of getting two Heads is the same as the probability of getting two Tails, namely, $\frac{1}{2} \times \frac{1}{2} = \frac{1}{4}$. The probability of both (dice and coins) is the product of the individual probabilities: $\frac{1}{4} \times \frac{1}{36} = \frac{1}{104}$. The probability of getting one or the other is the sum of the probabilities: $\frac{1}{4} + \frac{1}{36} = \frac{40}{104} = \frac{5}{13}$.

• The following sum has two missing terms. What is the median of the five numbers?

5, ___, 4, 1, ___

A. 10

B. 5

C. 9

D. 3

E. Not enough data to solve the problem

• • Answer E. To find the median, one must place the numbers in ascending order which is not possible if numbers are missing.

• ♥ •

• For the three factorial expressions 0! ; $(n-n)!$; and 1!

A. 0! and $(n-n)!$ are equivalent

B. $(n-n)!$ and 1! are equivalent

C. 0! and 1! are equivalent

D. All are equivalent

E. None of the above

• • Answer D.

$$0! = 1;\ (n-n)! = 0! = 1;\ 1! = 1.$$

• The following sum has two missing terms. What is the mean (average) of the five numbers? $5 + 4 + 1 + \_\_\_ + \_\_\_ = 25$

A. 10

B. 5

C. 9

D. Not enough data to solve the problem

E. None of the above

• • Answer B.

The mean is unaffected by the missing numbers because they already gave you the sum.

• ♥ •

• A wardrobe contains 3 scarves, 4 types of slacks, 4 blouses, and 3 belts and 3 pairs of shoes. If one item from each category is worn each day, how many days can you wear a different outfit each day?

A. A full year (365 days)

B. 100 days

C. More than a year (more than 365 days)

D. 144 days

E. None of the above

• • Answer C.

$$3 \cdot 4 \cdot 4 \cdot 3 \cdot 3 = 432$$

• The probability of rolling a seven with a single roll of a pair of dice is:

A. ½
B. ²⁄₇
C. ¹⁄₃₆
D. ¹⁄₇
E. None of the above

• • Answer E.

There are 6 × 6 = 36 possible outcomes from rolling a pair of dice.
There are 6 ways to roll a seven: 6&1 ; 5&2 ; 4&3 ; 1&6 ; 2&5; and 3&4 so there is a $\frac{6}{36} = \frac{1}{6}$ probability of rolling a seven with a pair of dice (the same probability of rolling a specified number on a single die).

• ♥ •

• The odds of winning a race are 3 to 1. What is the probability of losing the race?

A. 75%
B. 30%
C. 40%
D. 25%
E. None of the above

• • Answer D.

$$P_{win} = \frac{3}{4} = 0.75 = 75\%$$

$$P_{lose} = 1 - P_{win} = 1 - 0.75 = 0.25 = 25\%$$

• Given a sack of 12 marbles (4 black, 5 red, 3 white), the probability of getting a black marble, a red marble and a white marble (with replacement after each try) is

A. $\frac{1}{3}$
B. $\frac{5}{144}$
C. $\frac{2}{3}$
D. 1
E. None of the above

• • Answer B.
With replacement, the probability of picking a black marble is $\frac{4}{12} = \frac{1}{3}$; that of picking a red marble is $\frac{5}{12}$; and of picking a white marble is $\frac{3}{12} = \frac{1}{4}$. The probability of selecting all three is $\frac{1}{3} \times \frac{5}{12} \times \frac{1}{4} = \frac{5}{144}$

• ♥ •

• The probability of selecting a spade from a deck of 52 standard playing cards (no Joker) is

A. 20%
B. 52%
C. $\frac{1}{52}$
D. $\frac{1}{4}$
E. None of the above

• • Answer D.
There are 13 spades in a four-suit deck of playing cards, so the probability of selecting a spade is $\frac{13}{52} = \frac{1}{4}$

• The type study that involves testing a hypothesis is

A. observational study
B. experimental study of animals
C. frequency analysis of consumer spending
D. poll
E. None of the above

• • Answer B.
The experiment is used to test (accept or reject) a hypothesis.

• ♥ •

• A student's test score is exceeded only by 6% of those tested, and so is in the

A. 6th percentile
B. 50th percentile
C. 94th percentile
D. 100th percentile
E. None of the above

• • Answer C.
94% of those tested had a lower test score.

• A coin is tossed three times What is the probability that the result will be exactly two heads?

A. ½
B. ⅛
C. ¼
D. ⅜
E. None of the above

• • Answer D.

There are eight possible outcomes: HHH; HHT; HTH; HTT; THH; THT; TTH; and TTT.

Three of these outcomes have exactly two H's, so the probability is

(desired outcomes)/(total outcomes) = ⅜.

• ♥ •

• Analysis of a certain data set results in the following percentiles: Lower Quartile, Q_1, is 56; the Median, Q_2, is 65; and the Upper Quartile, Q_3, is 86. The range of the data set is 100, with the highest value equal to 110. The lowest value in the data set is

A. 30 and is the average
B. 10 and is an outlier
C. 56 and is not an outlier
D. unable to be calculated with the given data
E. None of the above

• • Answer B.

10 (highest value minus the range) is an outlier, determined as follows:

IQR = 86 – 56 = 30. 1.5×IQR = 1.5×30 = 45. Q_1 – 45 = 56 – 45 = 11.

The smallest value, 10, is an outlier because it is smaller than 11.

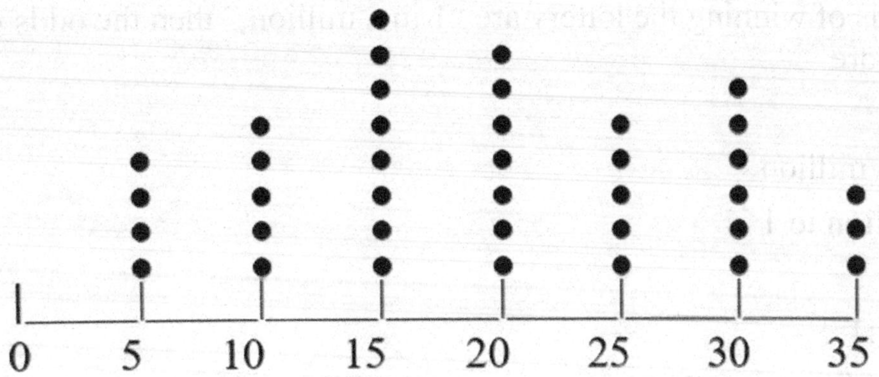

• The line plot show above displays the _____ of the data

A. mean

B. median

C. mode

D. standard deviation

E. None of the above

• • Answer C. The plot shows that the most frequent value is 15 and is the mode.

• In presentation of statistical data, a frequency polygon displays the

A. distribution of dependent variables over independent variables.

B. the median

C. the mean

D. data as a stacked bar chart

E. None of the above

• • Answer A.

The frequency displays the shape of the frequency distribution and is a graph of occurrence (dependent values) vs values (independent values) of a data set.

• If the odds of winning the lottery are "1 in a million," then the odds of losing the lottery are

A. 1 to 1 million
B. 1 million to 1
C. even
D. 1,000:1
E. None of the above

• • Answer B. Just "flip" the odds

• ♥ •

• Present the following data in a stem-and-leaf plot: 63, 12, 72, 80, 81, 90, 99, 3, 1

• • Answer

STEM	LEAF
0	1 3
1	2
6	3
7	2
8	0 1
9	0 9

• Everyone in a population is interviewed. This is an example of a

A. census
B. poll
C. survey
D. All of the above
E. None of the above

• • Answer A.

The census involves every member of a population

• ♥ •

• A way to organize and analyze statistical data involving two categories is through use of the

A. scatter plot
B. two-way frequency table.
C. three-way frequency table
D. any of the above
E. None of the above

• • Answer B.

The two-way frequency table displays data as percentage of column totals, row totals, or grand totals.

• The management staff of a major corporation arranged for the performance of a large number of statistical tests, then reported only the results of tests that provided favorable results. This is an example of

A. Flawed correlation
B. Selective bias
C. Faulty polling
D. Data fishing
E. Standard deviation

• •Answer D.

This is a statistical fallacy called data fishing (also called "data dredging," "data snooping," and other terms)

• ♥ •

• A guilty person is acquitted of a crime. This is an example of a

A. Type I error
B. Type II error
C. False positive
D. Selective bias
E. None of the above

• • Answer B.

A Type II error, or false negative.

• A company wants to present its salary structure in the best light to a group of prospective employees. Which statistic gives the highest "average salary?"

Position	Annual Salary
CEO	$400,000
CFO	350,000
Office Manager	100,000
Supervisor	80,000
Analyst 1	65,000
Analyst 2	35,000
Analyst 3	35,000
Trainee	20,000

A. range

B. mode

C. mean

D. median

E. All give the same result

• • Answer C.

The mean salary is $135,625; the median, $72,500, and the mode, $35,000. The range is inappropriate to the question.

• ♥ •

• Seven men selected at random had weights of 156, 165, 172, 195, 205, 220 and 200 lbs. How far from the *mean* was the weight of the heaviest man? How far from the *median* was the lightest man? Round the weight values to the nearest pound.

A. 15 and 15

B. 32 and 29

C. 39 and 22

D. 15 and 39

E. None of the above

• • Answer E.

The mean is $(156 + 165 + 172 + 195 + 205 + 220 + 200)/7 = {}^{1313}\!/_{7} = 187.6$ rounded up to 188. The heaviest man is $220 - 188 = 32$ lbs away from the mean; the lightest man is $195 - 156 = 39$ lbs away from the median.

• A sack of marbles holds 12 red, 5 green, 13 white, 8 black and 10 blue. Select two marbles without replacement; what is the probability that the both marbles are blue? Round intermediate calculations to four decimal places; round the final answer to two decimal places and express your answer in percent.

A. 21%
B. 2%
C. 3%
D. 4%
E. None of the above

• • Answer D.

The sack contains 48 marbles, ten of which are blue. The probability of selecting a blue marble on the first try is $^{10}/_{48} \approx 0.2083$. With 47 marbles remaining, the probability of again selecting a blue marble is $^{9}/_{47} \approx 0.1915$. The probability of two successful outcomes is the product of the probabilities of each: $0.2083 \times 0.1915 = 0.0399$ or 4% after rounding.

• ♥ •

• The median of the dataset {0, 1, 4, 5, 11, 13, 13, 13} is

A. 11
B. 12
C. 13
D. 16
E. None of the above

• • Answer E.

The median of a set containing an even number of members is the mean of the two middle numbers, in this case, $(5 + 11)/2 = 8$.

• The results of a student test show that the mean score is 76 with a standard deviation of 3 points. Assuming a normal distribution of scores, which is the best interpretation of these results?

A. The median score is not 76

B. Half of the test scores are greater than 79

C. Most of the test scores lie between 73 and 79

D. The lowest test score is 70 and the highest 82

E. None of the above are correct

• • Answer C.

A little over 68% of the test scores lie within one standard deviation from the mean (assuming a normal distribution). In connection with answer A, for a normal distribution, the median, mode, and mean are equal.

• ♥ •

• A toothpaste manufacturer launches a survey in a large city regarding consumer attitudes toward a particular brand of toothpaste. Which of the following groups would be the best survey sample?

A. Customers who have purchased the brand

B. Random selection from voter registration records

C. City residents selected using a random process

D. Calling every tenth entry listed in the city telephone directory.

E. None of the above are suitable.

• • Answer C.

The other choices are not true representative samples of a city population.

• A sack contains 200 marbles of various colors: 10 red; 24 black; 22 white; 13 blue; 16 yellow; 6 green; and 100 clear (no color). A child reaches into the sack and removes three marbles. What is the probability that all three marbles are clear?

A. $\frac{1}{2} \cdot \frac{1}{2} \cdot \frac{1}{2} = \frac{1}{8} = 0.125 = 12.5\%$
B. $\frac{1}{2} \cdot \frac{99}{200} \cdot \frac{98}{200} = \frac{9702}{80{,}000} = 0.121275 \approx 12.1\%$
C. $\frac{1}{2} \cdot \frac{99}{199} \cdot \frac{98}{198} = \frac{9702}{78{,}804} = 0.06156 \approx 12.3\%$
D. $\frac{3}{200} = 0.015 \approx 1.5\%$
E. None of the above

• • Answer C
This is a "without replacement" problem where the contents of the sack are reduced by one after each selection. The first three answers are nearly equal; the point of this question is that for this problem the colors of the other marbles are not relevant – only the total number in the sack.

• ♥ •

• Flip a coin and roll a pair of dice; how large is the sample space (i.e., possible outcomes) for this experiment?

A. 25
B. 24
C. 36
D. 72
E. None of the above

• • Answer D.
Two for the coin (H, T), six for the first die (1,2,3,4,5,6) and six for the second die (1,2,3,4,5,6). There are $2 \cdot 6 \cdot 6 = 72$ possible outcomes.

- The interquartile range is calculated as
    - A. $Q_4 - Q_3$
    - B. Q_3 – median
    - C. $Q_3 - Q_2$
    - D. $Q_3 - Q_1$
    - E. Highest – Lowest value

• • Answer D.

Note: Answers B and C are equivalent.

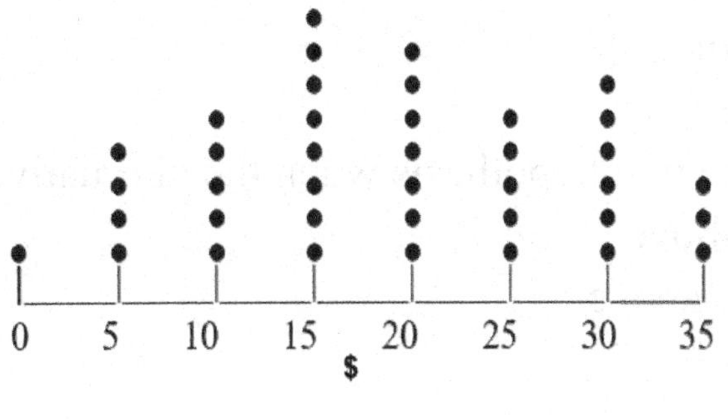

- In the line plot (or dot plot) shown above, the mode is
    - A. not displayed
    - B. 0
    - C. 35
    - D. 15
    - E. 10 and 25 (two modes)

• • Answer D.

• In an experiment you observe a difference when in fact there is none. This is an example of

A. a Type I error
B. a false positive
C. rejecting a null hypothesis when it is in fact true
D. All of the above
E. None of the above

• • Answer D.

• ♥ •

• Failing to observe an actual difference in a statistical test is an example of a

A. Type II error
B. false negative
C. rejecting the null hypothesis when the alternative hypothesis is true
D. All of the above
E. None of the above

• • Answer D.

• When examining a scatterplot, the action most likely to result in error is to

A. extrapolate the data to effect a prediction
B. interpolate between two adjacent data points
C. ignore an outlier
D. fit a line or a curve to the data points
E. None of the above will result in error

• • Answer A.

• ♥ •

• How should the analyst handle outliers in an experiment?

A. Ignore them
B. Determine the reason for the outlier
C. Average the outlier with the mean of the other data points
D. Redo the experiment by gathering new data

• • Answer B.
Determine if the outlier is due to math error or reporting error, for example.

• A correlation of -0.9 is

A. a near perfect fit
B. a poor fit
C. impossible to achieve
D. a standard deviation
E. None of the above

• • Answer A.

A correlation close to 1 or -1 (positive or negative correlation) is a near perfect fit.

• ♥ •

• Identify the joint frequencies in the tw0-way table shown below.

A. 55, 133, 188
B. 102, 68, 170
C. 55, 133, 102, 68
D. 157, 201, 188, 170
E. None of the above

WILL YOU PURSUE SCIENCE OR LIBERAL ARTS IN YOUR UPCOMING COLLEGE WORK?

	SCIENCE	ARTS	TOTAL
GIRLS	55	133	188
BOYS	102	68	170
TOTALS	157	201	358

• • Answer C.

THE MATH TUTOR'S FRIEND

INFORMAL REVIEW OF HIGH-SCHOOL MATH TOPICS FOR TUTORS

-5-

• Probability & Statistics •

Made in the USA
Monee, IL
02 November 2026

67382566R00044